Keri Rhinehart

ALGEBRA 2
Unit 1 Algebra 1 Revisited
5 Minute Math
5minmath.com

Rhinehart LLC

Cumming, GA 30040
5minmath.com

Copyright © 2021 by Rhinehart LLC. All rights reserved.

No part of this publication may be reproduced, stored in a retrieval system, or transmitted in any form or by any means, electronic, mechanical, photocopying, recording or otherwise, without prior written consent of the publisher.

ISBN: 9798456209078

Visit the Website

5minmath.com

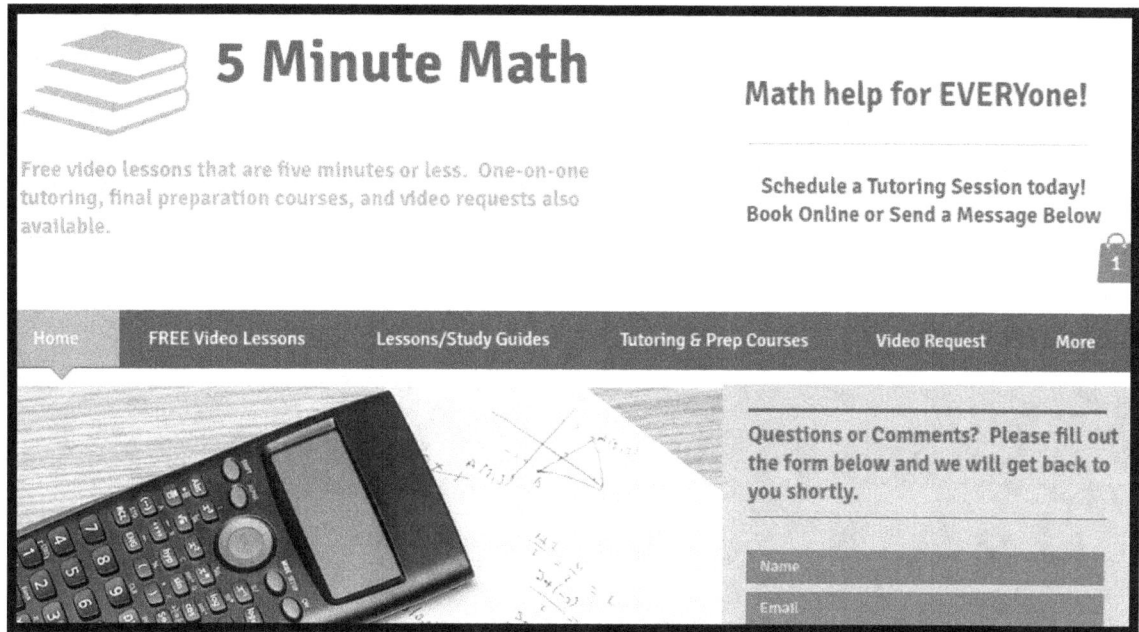

Free Lesson Videos, PDF Lessons, Tutoring, etc

Want Step by Step Solutions? Visit the website to purchase.

TABLE OF CONTENTS

Unit 1: Algebra 1 Revisited (Quadratics) Standards	7
Lesson 1: Literal Equations	8
Lesson 2: Complex Numbers	10
Lesson 3: Factoring Special Products	14
Quiz 1: Lessons 1 - 3	18
Lesson 4: Review Solving Quadratic Equations	19
Lesson 5: Expanding Using the Binomial Theorem	25
Algebra 1 Revisited - Test	27
Formulas	29
Calculator Guide	30
Solutions	31

Unit 1: Algebra 1 Revisited (Quadratics)

Literal Equations - Isolate a specific variable in a given equation.
Simplifying radicals and complex numbers – Simplify square roots by applying the product properties of radicals.
Complex Numbers - Identify square roots of negative numbers as imaginary and simplify them using properties of imaginary numbers.
Operations with Complex Numbers - Add, subtract, multiply and divide expressions involving complex numbers.
Factoring Special Products - Factor special polynomial cases, including Sum and Difference of Two Squares, Perfect Square Trinomial, Sum and Difference of Cubes.
Solving quadratic equations – Solve quadratic equations by factoring, taking the square root, completing the square and quadratic formula. Understand that the solution(s) to a quadratic equation represents the x-intercept(s)
Expanding using the Binomial Theorem - Expand the product of a binomial of degree 3 or higher using the Binomial Theorem.

Lesson 1: Literal Equations
Lesson 2: Complex Numbers
Lesson 3: Factoring Special Products
Quiz 1
Lesson 4: Review Solving Quadratic Equations
Lesson 5: Expanding Using the Binomial Theorem
Test

Unit 1 Algebra 1 Revisited - Lesson 1: Literal Equations

In this section, you will be asked to solve a given equation for a specified variable.

The variable you are asked to solve for, is the one that should be isolated. That means you need to move everything that is on the side with that variable to the other side.

Example 1: Solve $A = b + r$ for b

$A = b + r$
$-r -r$
$A - r = b$

$b = A - r$

Example 2: Solve $A = s^2$ for s

$A = s^2$
$\sqrt{A} = \sqrt{x^2}$
$\pm\sqrt{A} = s$

$s = \pm\sqrt{A}$

Example 3: Solve $y = mx + b$ for m

$y = mx + b$
$-b -b$
$y - b = mx$
$\div x \div x$
$\frac{y-b}{x} = m$

$m = \frac{y-b}{x}$

Example 4: Solve $y = \sqrt{x + 2}$

$y = \sqrt{x + 2}$
$y^2 = (\sqrt{x+2})^2$
$y^2 = x + 2$
$y^2 - 2 = x$

$x = y^2 - 2$

Example 5: Solve $F = \frac{x+y}{3}$ for y

$F = \frac{x+y}{3}$
$\times 3 \times 3$
$3F = x + y$
$-x -x$
$3F - x = y$

$y = 3F - x$

Practice

Solve the given equation for the indicated variable.

1. $h = s + t$, for s

2. $P = 4\sqrt{A}$, for A

3. $D = \frac{2x^2 - m}{3}$, for x

4. $F = \frac{9}{5}C + 32$, for C

5. $y = mx + b$, for m

6. $P = 2l + 2w$, for w

7. $y = k^2 - r$, for k

8. $M = \sqrt{\frac{x-1}{4}}$, for x

Unit 1 Algebra 1 Revisited - Lesson 2: Complex Numbers

Imaginary Number: A number that is not real. A number that when multiplied by itself (squared) is negative.

Imaginary numbers are helpful when we need to simplify that square root of a negative. Before, we could not simplify the square root of a negative, because there are no real numbers whose square is negative.

$$i = \sqrt{-1}$$

Powers of i:

$i = \sqrt{-1}$
$i^2 = -1$
$i^3 = -i$
$i^4 = 1$

Example 1: Simplify $\sqrt{-4}$

$\sqrt{-4} = \sqrt{-1 \cdot 4} = \sqrt{-1} \cdot \sqrt{4} = i \cdot 2 = \quad 2i$

Example 2: Simplify $2i\sqrt{-12x^3}$

$2i\sqrt{-12x^3} = 2i \cdot \sqrt{-1 \cdot 12 \cdot x^3} = 2i \cdot \sqrt{-1} \cdot \sqrt{12} \cdot \sqrt{x^3} = 2i \cdot i \cdot 2\sqrt{3} \cdot x\sqrt{x} = \quad 4ix\sqrt{3x}$

Example 3: Simplify the power of i. $-3i^{206}$

$-3i^{206} = -3(i^4)^{56}(i^2) = 3(1)^{56}(-1) = 3(1)(-1) = \quad -3$

Complex Number: An algebraic expression that includes a real number and an imaginary number separated by addition or subtraction. (ie. 2 + 7i)

Adding and Subtracting Complex Numbers

When adding and subtracting complex numbers, combine real numbers and combine imaginary numbers. Then simplify if possible.

Example 4: Simplify the following complex expression. Reduce any imaginary number to a power of 1.

$$12i - 7i$$
$$5i$$

Example 5: Simplify the following complex expression. Reduce any imaginary number to a power of 1.

$$(-2 + 3i) - (4 + 5i)$$
$$(-2 + 3i) + (-4 - 5i)$$
$$-2 - 4 + 3i - 5i$$
$$-6 - 2i$$

Multiplying Complex Numbers

When multiplying complex numbers, multiply coefficients and add exponents of i. Simplify if possible.

Example 6: Simplify the following complex expression. Reduce any imaginary number to a power of 1.

$$(2 + 3i)(3 - 5i)$$
$$2(3) + 2(-5i) + 3i(3) + 3i(-5i)$$
$$6 - 10i + 9i - 15i^2$$
$$6 - i - 15(-1)$$
$$6 - i + 15$$
$$21 - i$$

Conjugates: Two binomial expressions with the same values and opposite operation between the two values.
  Ie. 2 + 3i and 2 - 3i are conjugates

Example 7: State the conjugate of the given expression, then multiply the conjugates and simplify.

$$3 - 4i$$

The conjugate is $3 + 4i$

$$(3 - 4i)(3 + 4i)$$
$$3(3) + 3(4i) - 4i(3) - 4i(4i)$$
$$9 + 12i - 12i - 16i^2$$
$$9 - 16(-1)$$
$$9 + 16$$
$$25$$

Dividing Complex Numbers

When dividing complex numbers, divide coefficients and subtract powers of i. Simplify if possible.

Example 8: Simplify the following complex expression. Reduce any imaginary number to a power of 1.

$$\frac{48i^3}{6i}$$
$$8i^2$$
$$8(-1)$$
$$-8$$

Example 9: Simplify the following complex expression. Reduce any imaginary number to a power of 1.

$$\frac{18i^2 - 15i}{3i}$$
$$\frac{18i^2}{3i} - \frac{15i}{3i}$$
$$6i - 5$$

12

Rationalizing the denominator: When simplifying complex numbers, you cannot leave an imaginary number in the denominator. If there is an imaginary number in the denominator once you have simplified you must rationalize. To rationalize, you must multiply the numerator and denominator by:

1. If the denominator is a monomial: i
2. If the denominator is a binomial: The conjugate of the denominator.

What happens when you multiply complex conjugates?

   The imaginary numbers disappear

Example 10: Simplify the following complex expression. Reduce any imaginary number to a power of 1.

$\frac{4+i}{3i}$

$\frac{4+i}{3i} \times \frac{i}{i}$

$\frac{4i+i^2}{3i^2}$

$\frac{4i+(-1)}{3(-1)}$

$\frac{4i-1}{-3}$

$\frac{-4i+1}{3}$

$\frac{1-4i}{3}$

Example 11: Simplify the following complex expression. Reduce any imaginary number to a power of 1.

$\frac{2-3i}{5+6i}$

$\frac{2-3i}{5+6i} \times \frac{5-6i}{5-6i}$

$\frac{2(5)+2(-6i)-3i(5)-3i(-6i)}{5(5)+5(-6i)+6i(5)+6i(-6i)}$

$\frac{10-12i-15i+18i^2}{25-30i+30i-36i^2}$

$\frac{10-27i-18}{25+36}$

$\frac{-8-27i}{61}$

Practice

Simplify the following complex expressions. Reduce any imaginary number to a power of 1.

1. $16i - 8i$

2. $(3i)(5i^3)$

3. $(2i)(13i)$

4. $(4 + 3i) + (9 - 2i)$

5. $(15 - i) - (8 + 7i)$

6. $5(4 + 3i)$

7. $3i(8 - 2i)$

8. $(2 + 3i)(3 - 4i)$

9. $(5 - 6i)^2$

10. Multiply $(7 - 2i)$ by its conjugate

11. $\frac{21i}{7}$

12. $\frac{42i^3 + 15i}{3i}$

13. $\frac{2}{5i}$

14. $\frac{7}{4-i}$

15. $\frac{2+3i}{3-7i}$

16. $\sqrt{-4}$

17. $\sqrt{-8}$

18. $2\sqrt{-50x^2y^5}$

19. $-3i\sqrt{-6}$

20. i^{302}

21. $-15i^{675}$

Unit 1 Algebra 1 Revisited - Lesson 3: Factoring Special Products

*Notation to Understand: $\pm$ means keep the same sign, $\mp$ means switch the sign
*As always with factoring, be sure to take out the GCF first, if possible.

*If $a^2 = 16$ to find a, simply take the square root. $\sqrt{a^2} = \sqrt{16}$, $a = 4$

Example 1: Find a.

a. $a^2 = x^2$
$a = x$

b. $a^2 = 100$
$a = 10$

c. $a^2 = 9k^{10}$
$a = 3k^5$

d. $a^3 = 8$
$a = 2$

e. $a^3 = m^6$
$a = m^2$

Difference of Two Squares (DOTS)

$$a^2 - b^2 = (a+b)(a-b)$$
---

ie. $2y^2 - 98$

Factor out GCF: $2(y^2 - 49)$

$y^2 - 49$ is a DOTS because y^2 is a perfect square and 49 is a perfect square, and there is a subtraction sign in between.

$a^2 = y^2 \quad b^2 = 49$
$a = y \quad b = 7$

Use the Formula: $a^2 - b^2 = (a+b)(a-b)$
$(y+7)(y-7)$

Sum of Two Squares

$$a^2 + b^2 = (a+bi)(a-bi)$$
---

ie. $m^8 + 36$

$m^8 + 36$ is a Sum of Two Squares because m^8 is a perfect square and 36 is a perfect square, and there is an addition sign in between.

$a^2 = m^8 \quad b^2 = 36$
$a = m^4 \quad b = 6$

Use the Formula: $a^2 + b^2 = (a+bi)(a-bi)$
$(m^4 + 6i)(m^4 - 6i)$

Perfect Square Trinomial

$$a^2 \pm 2ab + b^2 = (a \pm b)(a \pm b) \text{ or } (a \pm b)^2$$
---

ie. $3x^2 + 30x + 75$

Factor out GCF: $3(x^2 + 10x + 25)$

$x^2 + 10x + 25$ is a Perfect Square Trinomial

$a^2 = x^2 \quad b^2 = 25$
$a = x \quad b = 5 \quad 2ab = 10x$

Use the Formula: $a^2 \pm 2ab + b^2 = (a \pm b)(a \pm b)$ or $(a \pm b)^2$

$3(x+5)(+5)$ or $3(x+5)^2$

Sum and Difference of Cubes

$$a^3 \pm b^3 = (a \pm b)(a^2 \mp ab + b^2)$$

Remember SOAP (Same, Opposite, Always Positive)

---

ie. $8x^3 - 1$

$8x^3 - 1$ is a Difference of Cubes

$a^3 = 8x^3 \quad b^3 = 1$
$a = 2x \quad b = 1 \quad ab = 2x$

Use the Formula: $a^3 \pm b^3 = (a \pm b)(a^2 \mp ab + b^2)$

$(2x - 1)((2x)^2 + 2x + (1)^2)$
$(2x - 1)(4x^2 + 2x + 1)$

Example 2: State what type of special product the following are. If not a special product, say so. Then factor if it is a special product.

a. $x^2 - 10$
   Not a special product. Many students think this is a DOTS, but it is not because 10 is not a perfect square.

b. $p^2 + 6p + 9$
   Perfect Square Trinomial
   $a^2 \pm 2ab + b^2 = (a \pm b)(a \pm b)$ or $(a \pm b)^2$
   $a^2 = p^2 \quad b^2 = 9$
   $a = p \quad b = 3 \quad 2ab = 6p$
   $(p + 3)(p + 3)$ or $(p + 3)^2$

   Check: $(p + 3)^2$
   $(p + 3)(p + 3)$
   $p^2 + 3p + 3p + 9$
   $p^2 + 6p + 9$ ✓

c. $u^3 - 27$
   Difference of Cubes
   $a^3 \pm b^3 = (a \pm b)(a^2 \mp ab + b^2)$
   $a^3 = u^3 \quad b^3 = 27$
   $a = u \quad b = 3$
   $(u - 3)(u^2 + 3u + 9)$

   Check: $(u - 3)(u^2 + 3u + 9)$
   $u^3 + 3u^2 + 9u - 3u^2 - 9u - 27$
   $u^3 - 27$ ✓

d. $4m^4 - 49$
   Difference of Two Squares
   $a^2 - b^2 = (a + b)(a - b)$
   $a^2 = 4m^4 \quad b^2 = 49$
   $a = 2m^2 \quad b = 7$
   $(2m^2 + 7)(2m^2 - 7)$

   Check: $(2m^2 + 7)(2m^2 - 7)$
   $4m^4 - 14m^2 + 14m^2 - 49$
   $4m^4 - 49$ ✓

e. $2y^2 + 8$
   Sum of Squares **(Must take out GCF first)**

$a^2 + b^2 = (a + bi)(a - bi)$
$2(y^2 + 4)$
$a^2 = y^2 \quad b^2 = 4$
$a = y \quad\quad b = 2$
$(y + 2i)(y - 2i)$
$2(y + 2i)(y - 2i)$

Check:
$2(y + 2i)(y - 2i)$
$2(y^2 - 2iy + 2iy - 4i^2)$
$2(y^2 - 4i^2)$
$2(y^2 - 4(-1))$
$2(y^2 + 4)$
$2y^2 + 8$ ✓

f. $9k^2 - 12k + 4$

Perfect Square Trinomial
$a^2 \pm 2ab + b^2 = (a \pm b)(a \pm b) \text{ or } (a \pm b)^2$
$a^2 = 9k^2 \quad b^2 = 4$
$a = 3k \quad b = 2 \quad 2ab = 12k$
$(3k - 2)(3k - 2) \text{ or } (3k - 2)^2$

Check:
$(3k - 2)(3k - 2) \text{ or } (3k - 2)^2$
$9k^2 - 6k - 6k + 4$
$9k^2 - 12k + 4$ ✓

Practice:

Factor the following. If not factorable, state so.

1. $k^2 + 9$

2. $2x^2 - 2$

3. $m^2 - 8x + 16$

4. $y^3 + 27$

5. $8b^3 - 81$

6. $4x^2 + 40x + 100$

7. $2m^3 - 242$

8. $6k^8 - 6$

9. $p^2 + 3$

10. $4u^2 - 12uv + 9v^2$

11. $m^3 - 343n^6$

12. $x^2 + 2x + 4$

Unit 1 Algebra 1 Revisited - Quiz 1: Lessons 1 - 3

Solve the following for the given variable

1. $A = P(1+r)^t$; r

2. $y = \sqrt{x-3}$; x

3. $m = \frac{b-p}{c}$, p

Simplify Completely

4. $(4+7i)^2$

5. $\sqrt{-50x^5}$

6. $\frac{3}{4i}$

7. $2i\sqrt{-50}$

8. $\frac{5}{2-i}$

9. $(5+11i) - (6-17i)$

Factor or state not factorable

10. $64m^3 - 27$

11. $x^2 + 14x + 49$

12. $5y^4 - 20$

13. $25p^2 - 60pm + 36m^2$

14. $7x^2 + 28$

15. $y^3 + 8k^6$

Unit 1 Algebra 1 Revisited - Lesson 4: Review Solving Quadratic Equations

Quadratic: A quadratic equation is an equation where there is a term where x is squared. Can be written in the form $ax^2 + bx + c = 0$, where a, b and c are integers.

<u>1. Solving by FACTORING</u>

Zero-Product Property: The zero product property states that if a·b=0 then either a or b equal zero. This basic property helps us solve equations like (2x-3)(x+4)=0

Example 1: Solve the following quadratic equation by factoring.

$(x + 3)(x - 5) = 0$
$x + 3 = 0 \qquad\qquad x - 5 = 0$
$x = -3 \qquad\qquad x = 5$

Example 2: Solve the following quadratic equation by factoring.

$4x(2x - 1) = 0$
$4x = 0 \qquad\qquad 2x - 1 = 0$
$x = 0 \qquad\qquad\quad 2x = 1$
$\qquad\qquad\qquad\qquad x = \frac{1}{2}$

Example 3: Solve the following quadratic equation by factoring.

$x^2 - 14x + 40 = 0$
$(x - 10)(x - 4) = 0$
$x - 10 = 0 \qquad\qquad x - 4 = 0$
$x = 10 \qquad\qquad\quad x = 4$

Example 4: Solve the following quadratic equation by factoring.

$4x^2 + 7x = -3$
$4x^2 + 7x + 3 = 0$
$(4x + 3)(x + 1) = 0$
$4x + 3 = 0 \qquad\qquad x + 1 = 0$
$4x = 3 \qquad\qquad\quad x = -1$
$x = \frac{3}{4}$

<u>2. Solve by taking the SQUARE ROOT</u>

This method should only be used in the following two situations:
1. The value of b in $ax^2 + bx + c$ is 0. This means there is a squared quantity and a constant. (ie. $5x^2 + 3 = 10$)
2. The quadratic is in vertex form: $a(x - h)^2 + k = 0$. (ie. $3(x - 1)^2 = 9$)

Example 1: Solve the following quadratic equation by taking the square root. If it cannot be solved by taking the square root, then state that.

$5x^2 = 10$
$x^2 = 2$
$x = \pm\sqrt{2}$

*Keep in mind that $\pm\sqrt{2}$ is two solutions. It represents $\sqrt{2}$ and $-\sqrt{2}$

Example 2: Solve the following quadratic equation by taking the square root. If it cannot be solved by taking the square root, then state that.

20
$3x^2 - 2 = 9$
$3x^2 = 11$
$x^2 = \frac{11}{3}$
$x = \pm\sqrt{\frac{11}{3}}$
$x = \pm\frac{\sqrt{11}}{\sqrt{3}}$
$x = \pm\frac{\sqrt{11}}{\sqrt{3}} \times \frac{\sqrt{3}}{\sqrt{3}} = \pm\frac{\sqrt{33}}{\sqrt{9}} = \pm\frac{\sqrt{33}}{3}$
$x = \pm\frac{\sqrt{33}}{3}$

Example 3: Solve the following quadratic equation by taking the square root. If it cannot be solved by taking the square root, then state that.

$(2x - 5)^2 = -81$
$\sqrt{(2x-5)^2} = \pm\sqrt{-81}$
$2x - 5 = \pm 9i$
$2x = 5 \pm 9i$
$x = \frac{5 \pm 9i}{2}$

Example 4: Solve the following quadratic equation by taking the square root. If it cannot be solved by taking the square root, then state that.

$3(x + 1)^2 - 10 = 2$
$3(x + 1)^2 = 12$
$(x + 1)^2 = 4$
$\sqrt{(x+1)^2} = \pm\sqrt{4}$
$x + 1 = \pm 2$
$x = -1 \pm 2$
$x = -1 + 2 \qquad\qquad x = -1 - 2$
$x = 1 \qquad\qquad\qquad x = -3$

3. Solve by COMPLETING THE SQUARE

When completing the square, you are finding the c value that can be added to $ax^2 + bx$ so that it can be factored into a perfect square (ie (x+2)(x+2) = $(x + 2)^2$)

To find the value that completes the square, take half of b, then square it.

Example 1: Find the value, c, that completes the square

$x^2 - 8x + c$

b = -8

Half of -8 is -4 ($-8 \div 2 = -4$)

-4 squared is 16 ($(-4)^2 = 16$)

Therefore, the value that completes the square is 16: $x^2 - 8x + 16$

Example 2: Find the value, c, that completes the square.

$x^2 + x + c$

b = 1

Half of 1 is $\frac{1}{2}$

$\frac{1}{2}$ squared is $\frac{1}{4}$

Therefore, the value that completes the square is $\frac{1}{4}$. $x^2 + x + \frac{1}{4}$

Once you find the value that completes the square, you will need to factor it into a perfect square binomial.

Example 3: Find the value, c, that completes the square, then factor it into a perfect square binomial.

$x^2 + 4x + c$

b = 4, half of 4 is 2, 2 squared is 4.

$x^2 + 4x + 4$

$(x+2)(x+2) = (x+2)^2$

Example 4: Find the value, c, that completes the square, then factor it into a perfect square binomial.

$x^2 - 3x + c$

b = -3, half of -3 is $\frac{-3}{2}$, $\frac{-3}{2}$ squared is $\frac{9}{4}$

$x^2 - 3x + \frac{9}{4}$

$(x - \frac{3}{2})(x - \frac{3}{2}) = (x - \frac{3}{2})^2$

To solve a quadratic equation by completing the square:
1. Isolate $ax^2 + bx$ on one side and the constant on the other.
2. Find the value that completes the square, then add it to both sides.
3. Factor the quadratic into a perfect square binomial.
4. Solve the quadratic by taking the square root.

Example 5: Solve the quadratic by completing the square.

$x^2 + 8x = 8$

$x^2 + 8x + 16 = 8 + 16$

$(x+4)^2 = 24$

$\sqrt{(x+4)^2} = \pm\sqrt{24}$

$x + 4 = \pm 2\sqrt{6}$

$x = -4 \pm 2\sqrt{6}$

Example 6: Solve the quadratic by completing the square.

$x^2 - 2x - 3 = 2$

$x^2 - 2x = 5$

$x^2 - 2x + 1 = 5 + 1$

$(x-1)^2 = 6$

22

$\sqrt{(x-1)^2} = \pm\sqrt{6}$

$x - 1 = \pm\sqrt{6}$

$x = 1 \pm \sqrt{6}$

4. Solve using the QUADRATIC FORMULA

The final way that we will look at solving quadratic equations is by using the **Quadratic Formula**. This formula can <u>ALWAYS</u> be used to solve any quadratic equation.

Quadratic Formula

$$x = \frac{-b \pm \sqrt{b^2 - 4ac}}{2a}$$

*Be sure that your quadratic equation is in the form $ax^2 + bx + c = 0$ before using the quadratic formula.

<u>To solve a quadratic equation by using the Quadratic Formula:</u>
1. Get quadratic equation into the form $ax^2 + bx + c = 0$
2. Write down the value of a, b and c
3. Plug a, b and c into the quadratic formula
4. Simplify

Example 1: Solve the following quadratic equation using the Quadratic Formula

$x^2 + 7x - 10 = 0$

a = 1, b = 7, c = -10

$\frac{-7 \pm \sqrt{(7)^2 - 4(1)(-10)}}{2(1)}$

$\frac{-7 \pm \sqrt{49 + 40}}{2}$

$\frac{-7 \pm \sqrt{89}}{2}$

Therefore, $x = \frac{-7 \pm \sqrt{89}}{2}$, which could also be written as $\frac{-7 - \sqrt{89}}{2}$ or $\frac{-7 + \sqrt{89}}{2}$

Example 2: Solve the following quadratic equation using the Quadratic Formula

$2x^2 - 6x = -13$

$2x^2 - 6x + 13 = 0$

a=2, b=-6, c = 13

$\frac{6 \pm \sqrt{(-6)^2 - 4(2)(13)}}{2(2)}$

$\frac{6 \pm \sqrt{36 - 104}}{4}$

$\frac{6 \pm \sqrt{-68}}{4}$

$\frac{6 \pm 2i\sqrt{17}}{4}$

$\frac{3 \pm i\sqrt{17}}{2}$

Therefore, $x = \frac{3 \pm \sqrt{17}}{2}$, which could also be written as $\frac{3+\sqrt{17}}{2}$ or $\frac{3-\sqrt{17}}{2}$

Example 3: Solve the following quadratic equation using the Quadratic Formula

$-3x^2 + 4x = x^2 - 3$

$-4x^2 + 4x + 3 = 0$
a = -4, b = 4, c = 3

$$\frac{-4 \pm \sqrt{(4)^2 - 4(-4)(3)}}{2(-4)}$$
$$\frac{-4 \pm \sqrt{16 + 48}}{-8}$$
$$\frac{-4 \pm \sqrt{64}}{-8}$$
$$\frac{-4 \pm 8}{-8}$$
$$\frac{-4+8}{-8} \text{ or } \frac{-4-8}{-8}$$
$$\frac{4}{-8} \text{ or } \frac{-12}{-8}$$

Therefore, $x = -\frac{1}{2}$ or $\frac{3}{2}$

Practice

Solve the following quadratic equations using the most appropriate method.

1. $x^2 - 7x = 8$

2. $4x^2 - 11 = 5$

3. $2x^2 - 8 = 3x^2 + 4x$

4. $13 = 3(x-4)^2 + 1$

5. $5x^2 + 3x - 1 = 0$

Unit 1 Algebra 1 Revisited - Lesson 5: Expanding Using the Binomial Theorem

How would you expand the following $(x-3)^6$?

$(x-3)(x-3)(x-3)(x-3)(x-3)(x-3)$
$(x^2 - 3x - 3x + 9)(x-3)(x-3)(x-3)(x-3)$
$(x^2 - 3x + 9)(x-3)(x-3)(x-3)(x-3)$
$(x^3 - 3x^2 - 3x^2 + 9x + 9x - 27)(x-3)(x-3)(x-3)$
$(x^3 - 6x^2 + 18x - 27)(x-3)(x-3)(x-3)$
. . . . . .
. . . . . .
. . . . . .
. . . . . .

This would take a very long time to expand. However, there is a method to expand binomials that is much more efficient.

Binomial Theorem: A method used to expand binomials using Pascal's Triangle.

Pascal's Triangle

```
              1
             1 1
            1 2 1
           1 3 3 1
          1 4 6 4 1
        1 5 10 10 5 1
       1 6 15 20 15 6 1
```

Each row is found by adding the two values above it and placing a 1 on each end.

Steps for Expanding using the Binomial Theorem:
1. List the numbers from Pascal's triangle vertically (use the row with the second value that is the same as your exponent).
2. List the first term from the binomial next to each value from Pascal's Triangle.
3. Raise the first term in the first row to the power of the binomial. The second row by one less, the third row by one less, etc. (In the last row, the first term should be raised to a power of zero)
4. List the second term from the binomial in each row.
5. Raise the second term in the first row to the power or zero. The second row by one more, the third row by one more, etc. (In the last row, the second term should be raised to the power of the binomial)

Example 1: Expand $(m+4)^3$ using the Binomial Theorem

$1(m)^3(4)^0 = 1(m^3)(1) = m^3$
$3(m)^2(4)^1 = 3(m^2)(4) = 12m^2$
$3(m)^1(4)^2 = 3(m)(16) = 48m$
$1(m)^0(4)^3 = 1(1)(64) = 64$

$m^3 + 12m^2 + 48m + 64$

Example 2: Expand $(2x-3)^5$ using the Binomial Theorem

$1(2x)^5(-3)^0 = 1(32x^5)(1) = 32x^5$
$5(2x)^4(-3)^1 = 5(16x^4)(-3) = -240x^4$
$10(2x)^3(-3)^2 = 10(8x^3)(9) = 720x^3$
$10(2x)^2(-3)^3 = 10(4x^2)(-27) = -1080x^2$
$5(2x)^1(-3)^4 = 5(2x)(81) = 810x$
$1(2x)^0(-3)^5 = 1(1)(-243) = -243$

$32x^5 - 240x^4 + 720x^3 - 1080x^2 + 810x - 243$

Practice:

Write the first 10 rows of Pascal's Triangle

Expand using the Binomial Theorem:

1. $(y+5)^4$
2. $(m-2)^5$
3. $(2x+1)^6$
4. $(3y-4)^3$

State the 3rd term in the expansion of the binomial

5. $(x+8)^4$
6. $(4t-3)^7$

State the coefficient of x^3 in the expansion of the binomial

7. $(x+4)^8$
8. $(5x-2)^3$

Algebra 2 Unit 1 Algebra 1 Revisited - Test

Solve the following for the given variable

1. $x + 3m = y$; m

2. $M = 2\sqrt{p - t}$; p

3. $s = \frac{p^2}{5}$; p

Simplify Completely

4. $\sqrt{-98m^7}$

5. $\frac{4+7i}{3i}$

6. $(3 - 4i)(5 + 3i)$

7. $3i\sqrt{-25}$

8. $\frac{12}{5-7i}$

9. $(7 + 2i) - (12 - 9i)$

Factor or state not factorable

10. $p^4 - 4p^2t + 4t^2$

11. $125y^3 - 8$

12. $2x^2 - 32$

13. $2m^3 - 54p^3$

14. $x^2 + 24x + 144$

15. $u^4 + 16$

Solve the quadratic equation using any method

16. $x^2 - 12x = -35$

17. $3(y-3)^2 = 15$

18. $4m^2 - 7m = 3m + 5$

19. $p^2 - 9 = 0$

20. $3u^2 = 8u - 1$

21. $12 = 3 - 3(5x+4)^2$

Expand the following using the Binomial Theorem

22. $(x-2)^5$

23. $(4m + 3p)^4$

24. $(2y^3 - m)^3$

Formulas

Difference of Two Squares (DOTS)

$$a^2 - b^2 = (a+b)(a-b)$$

Sum of Two Squares

$$a^2 + b^2 = (a+bi)(a-bi)$$

Perfect Square Trinomial

$$a^2 \pm 2ab + b^2 = (a \pm b)(a \pm b) \text{ or } (a \pm b)^2$$

Sum and Difference of Cubes

$$a^3 \pm b^3 = (a \pm b)(a^2 \mp ab + b^2)$$

Remember SOAP (Same, Opposite, Always Positive)

Quadratic Formula

$$x = \frac{-b \pm \sqrt{b^2 - 4ac}}{2a}$$

Calculator Guide (TI-84)

Using a Calculator (TI-84 Plus) to get values for a function

Step 1: Press [Y=]
Step 2: Type your function into calculator
Step 3: Press [2ND] [WINDOW (TBLSET)]
Step 4: Enter the value where you want the table to start (TblStart), the amount you want the x value to increase each time (^Tbl). If you want automatic values choose Auto for Indpnt. If you want to input specific values for x, choose Ask for Indpnt. Make sure Depend is always Auto.
Step 5: Press [2ND] [GRAPH (TABLE)]. If you are entering specific values for x, type the value and press [ENTER]

Using a Calculator (TI-84 Plus) to graph a function

Step 1: Press [Y=]
Step 2: Type your function into calculator
Step 3: Press [ZOOM] 6

```
ZOOM        MEMORY
2: Zoom In
3: Zoom Out
4: ZDecimal
5: ZSquare
6: ZStandard
7: ZTrig
9: ZoomStat
```

Step 4: If you want to change your picture, press [WINDOW]. (Xmin is the left bound, Xmax is the right bound, Xscl is the amount between each x value, Ymin is the bottom bound, Ymax is the top bound, Yscl is the amount between each y value.
Step 5: Press [GRAPH]

Using the TRACE feature on the Calculator (TI-84 Plus)

Step 1: Graph your function
Step 2: Press [TRACE]
Step 3: Use the left and right cursor to scroll along the function. You will see the ordered pairs for each point at the bottom.
Step 4: To find specific values, suchs as zeros, minimums, etc, Press [2ND] [TRACE (CALC)]. Choose the value you want to find.
Step 5: Scroll to the left of the value you are trying to find, Press [ENTER]. Scroll to the right of the value you are trying to find, Press [ENTER]. Press [ENTER] again.

Unit 1 Solutions

Unit 1 Lesson 1

1. $s = h - t$
2. $A = \frac{p^2}{16}$
3. $x = \pm\sqrt{\frac{3D+m}{2}}$
4. $C = \frac{5}{9}(F - 32)$
5. $m = \frac{y-b}{x}$
6. $w = \frac{p-2l}{2}$
7. $k = \pm\sqrt{y + r}$
8. $x = 4m^2 + 1$

Unit 1 Lesson 2

1. $8i$
2. 15
3. -26
4. $13 + i$
5. $7 - 8i$
6. $20 + 15i$
7. $6 + 24i$
8. $18 + i$
9. $-11 - 60i$
10. 45
11. $3i$
12. -9
13. $-\frac{2i}{5}$
14. $\frac{28+7i}{17}$
15. $\frac{15+23i}{58}$
16. $2i$
17. $2i\sqrt{2}$
18. $10ixy^3\sqrt{2}$
19. $3\sqrt{6}$
20. -1
21. $15i$

Unit 1 Lesson 3:

1. $(k + 3i)(k - 3i)$
2. $2(x + 1)(x - 1)$
3. $(m - 4)^2$ or $(m - 4)(m - 4)$
4. $(y + 3)(y^2 - 3y + 9)$
5. Not Factorable
6. $4(x + 5)^2$ or $4(x + 5)(x + 5)$
7. $2(m^3 - 121)$
8. $6(k - 1)(k + 1)(k + i)(k - i)(k^2 + i)(k^2 - i)$
9. Not Factorable
10. $(2u - 3v)^2$
11. $(m - 7n^2)(m^2 + 7mn^2 + 49n^4)$
12. Not Factorable

Quiz 1

1. $r = \pm\sqrt{\frac{A}{p}} - 1$
2. $x = y^2 + 3$
3. $p = b - mc$
4. $-33 + 56i$
5. $5ix^2\sqrt{2x}$
6. $-\frac{3i}{4}$
7. $-10\sqrt{2}$
8. $2 + i$
9. $-1 + 28i$
10. $(4m - 3)(16m^2 + 12m + 9)$
11. $(x + 7)^2$
12. $5(y^2 - 2)(y^2 + 2)$
13. $(5p - 6m)^2$
14. $7(x + 2i)(x - 2i)$
15. $(y + 2k^2)(y^2 - 2yk^2 + 4k^4)$

Unit 1 Lesson 4

1. $x = 8, -1$
2. $x = \pm 2$
3. $x = -2 \pm 2i$
4. $x = 6, 2$
5. $x = \frac{-3\pm\sqrt{29}}{10}$

Unit 1 Lesson 5

```
         1
        1 1
       1 2 1
      1 3 3 1
     1 4 6 4 1
    1 5 10 10 5 1
   1 6 15 20 15 6 1
  1 7 21 35 35 21 7 1
 1 8 28 56 70 56 28 8 1
1 9 36 84 126 126 84 36 9 1
1 10 45 120 210 252 210 120 45 10 1
```

1. $y^4 + 20y^3 + 150y^2 + 500y + 625$
2. $m^5 - 10m^4 + 40m^3 - 80m^2 + 80m - 32$
3. $64x^6 + 192x^5 + 240x^4 + 160x^3 + 60x^2 + 12x +$
4. $27y^3 - 108y^2 + 144y - 64$
5. $384x^2$
6. $193536t^5$
7. 57344
8. 125

Unit 1 Test:

1. $m = \frac{y-3}{3}$
2. $p = \frac{rt}{4} + t$
3. $p = \pm\sqrt{55}$
4. $7im^3\sqrt{2m}$
5. $\frac{7-4i}{3}$
6. $27 - 11i$
7. -15
8. $\frac{30+42i}{37}$
9. $-5 + 11i$
10. $(p^2 - 2t)^2$
11. $(5y - 2)(25Y^2 + 10y + 4)$
12. $2(x - 4)(x + 4)$
13. $2(m - 3p)(m^2 + 3mp + 9p^2)$
14. $(x + 12)^2$
15. $(u^2 - 4i)(u^2 + 4i)$
16. $x = 5, 7$
17. $y = 3 \pm \sqrt{5}$
18. $x = \frac{5\pm 3\sqrt{5}}{4}$
19. $p = \pm 3$
20. $u = \frac{4\pm\sqrt{13}}{3}$
21. $x = \frac{-4\pm\sqrt{3}}{5}$
22. $x^5 - 10x^4 + 40x^3 - 80x^2 + 80x - 32$
23. $256m^4 + 768m^3p + 864m^2p^2 + 432mp^3 + 81p^4$
24. $8y^9 + 12y^6m + 6y^3m^2 + m^3$

www.ingramcontent.com/pod-product-compliance
Lightning Source LLC
Chambersburg PA
CBHW082124220526
45472CB00009B/2296